SWINDON WORKS

Images at and accessible to the Transport Treasury archive

200 Years of Train Travel

ISBN 978-1-917776-04-2

First published in 2025 by Transport Treasury Publishing Ltd. 16 Highworth Close, High Wycombe, HP13 7PJ
Totem Publishing, an imprint of Transport Treasury Publishing.

www.ttpublishing.co.uk

Printed in Tarxien, Malta by Gutenberg Press Ltd.

Front cover - To many the pinnacle of Swindon's achievements was the steam era - after all, it lasted for well over a century. In this early 1960s image it was rapidly drawing to a close, new steam engine construction having ceased in 1960 although overhauls and repairs would continue until about 1964. Sometimes the newly repaired engined would be destined for a short life, such was the pace of progress towards dieselisation. Here we have newly overhauled No 5071 *Spitfire* of the famed Castle class and renamed from *Clifford Castle* in 1940. Built in 1938, it would survive until October 1963 before being withdrawn, and uncermoniously tuned into baked bean cans, razor blades or whatever. Alongside is 2-8-0 No 3800, also built in 1938 and which would be working until August 1964.

Frontispiece - Some diesel products of Swindon had even shorter lives. This is one example, No D821 *Greyhound* of the Warship class. Built at Swindon in 1960 and withdrawn after just 12 years service in 1972, hardly an efficient retrun for the taxpaxer. *Greyhound* in seen by the turntable at Swindon Works in September 1977 fortunately having escaped the cutters torch and gaining the privilege of being the first main line diesel to be bought for preservation - many more would follow. The locomotive is maintained in working order and is based on the Severn Valley Railway. *(G H Taylor)*

Rear cover - Part of the historic railway village at Swindon; houses built for the skilled artisans in whose hands the famed products of Swindon were produced. Families and often even lodgers would be accommodated in these properties. The cottages themselves were designed by Matthew Digby Wyatt the architect of Paddington station. Bath stone was used which came from the excavating of Box Tunnel. The shareholders of the GWR also provided sufficient funds for a school, church and adult education centre, the intention being to provide for a sense of community. The Swindon railway viillage is one of the earliest examples of an 'industrial village'. *(Graham Smith)*

Copies of most of the images within this book are available direct from The Transport Treasury. Please quote the book title, page number and image reference.

INTRODUCTION

The Great Western Railway (GWR) and Swindon are inextricably linked, forming a vital chapter in both British railway history and the town's very identity. The arrival of the GWR in the 19th century transformed Swindon from a small market town into a thriving industrial hub. What follows is a deeper look into this significant relationship:

The Genesis of Swindon's Railway Heritage

The GWR, masterminded by the renowned engineer Isambard Kingdom Brunel, aimed to connect London with Bristol. The location of Swindon as a locomotive facility was strategically chosen due to its position roughly midway along the route, and also due to the change in gradients of the track; that to the east being a gentle fall over the 77 miles to Paddington - the top of St Paul's Cathedral was level with Swindon station. To the west however, there were (for the time) what were steep gradients as well as several tunnels including the famous Box Tunnel at 1.83 miles in length.

The original route had been planned to pass further south through Savernake Forest but this was on land owned by the Marquess of Ailesbury who had objected to the new form of transport.

At this stage the Great Western Railway operated on the broad gauge of 7' 0¼". It was eventually forced to capitulate to the standard 4' 8½" gauge by 1892 after which development continued in leaps and bounds under men of vision such as William Dean, George Jackson Churchward, Charles Collett and Frederick Hawksworth.

Swindon Works

In the 1840s, Swindon was selected as the site for the GWR's major locomotive construction and repair works. This decision had a profound impact, leading to the rapid growth of the town and the creation of a vast industrial complex.

Over time Swindon Works became a by-word for excellence producing some of the most iconic steam locomotives in history.

Swindon Railway Station

The railway station itself became a crucial point on the GWR network. Notably, it was famous (but as time passed for all the wrong reasons) for its refreshment rooms, a legal agreement meaning all trains passing through Swindon were required to stop for passengers to take a break on their journey.

For several decades, Swindon was the location where the engines of the trains would be changed due to the differing gradient either side, this adding to the station's importance. The importance of Swindon station had also increased in consequence of the opening of the line from Swindon through to Kemble and Gloucester in 1844.

The Impact on Swindon: Industrial Growth

The GWR Works provided employment for thousands of people, driving the town's population and economic expansion. Swindon became a quintessential railway town, with its social and cultural life deeply intertwined with the railway industry.

Community Development

The GWR played a significant role in developing the town's infrastructure, including housing, schools, and healthcare facilities. The company also fostered a strong sense of community among its employees. The railway village at Swindon, home to some of the GWR employees came about as there was no spare housing available in the town. A new village was ordered by the GWR Board from Mr Brunel – an estate of 300 'cottages' including a number of shops on the corners and a 'barracks' building for crowds of single workmen. An open field was left at the centre, lined with trees, which served as a 'promenade' until 1854 when it was built upon.

The village was complete by 1842, with St Mark's Church and a school built nearby by 1845.

Families lived crowded together in tiny houses, with an average of 7 (and as many as 12) people housed in each, sometimes including lodgers! In the beginning they had no running water or even toilets – only an 'outhouse' with a hole in the ground.

The Products of Swindon

The locomotives and rolling stock of the designers mentioned were produced in the locomotive, carriage and wagons shops of the site. Such was the expanse and expertise available that it is almost impossible to accurately determine the number of each produced; although engine classes alone could be in single figures or up to several hundred, with carriages built in batches to what were referred to as 'Diagrams' and each group as a 'Lot'. A similar policy was followed for wagons from the large 'Crocodile' type to the small 'Mink' and 'open' wagons. (The GWR had a telegraph code for each wagon type; an early form of shorthand intended to reduce the time taken to telegraph information and requirements.) Construction of new steam locomotives ceased in 1960 and diesels a few years later. After this the works was kept busy on repairs and overhauls but as time passed so it also became a place where redundant assets would accumulate.

A Lasting Legacy

Regretfully after a long fight British Rail made the decision to close the Works, but not before one former employee had commented, 'If a High Speed Train breaks down in the station we know how to fix it but we are not allowed to.'

Sadly too that great cathedral that was the epi-centre of Swindon Works itself, 'A Shop' would not survive the demolition contractors and was pulled down to be replaced with blocks of housing. Part of the works does survive to today as a modern shopping centre whilst the former Pattern Store is now the renowned STEAM Museum of the Great Western Railway. English Heritage has also moved into the former office block.

Although the works closed in 1986, its legacy lives on in the town's architecture, museums, and cultural identity.

Preservation, both officially through the National Collection and thanks to the ranks of enthusiasts and benefactors, has seen more than 150 former GWR locomotives survive although certainly not all are in running order at the same time. Gaps in history, particularly from the early years are also being slowly filled for the benefit of present and future generations.

Social Impact

The GWR's influence extended beyond the workplace, shaping the social fabric of Swindon. The company provided welfare and recreational facilities for its employees, contributing to their well-being.

Modern-Day Swindon

Swindon station continues to be an important station on the Great Western Main line. The town's connection to the railway is still very strong, and is a key part of its identity. Passenger numbers peaked pre-Covid at 3.7 million and after falling to just 0.65 million during the pandemic have recovered to 2.8 million in 2023/2024.

In essence, the Great Western Railway's presence in Swindon was transformative, leaving an indelible mark on the town's history and character. Today, Swindon stands as a testament to the golden age of steam and the enduring legacy of the GWR.

Not surprisingly there are no photographs of the construction of the GWR nearly two centuries ago and what does survive in the form of artist's woodcuts, good though they are, cannot provide the detail which we have now come to expect.

Fast forward then to the end of the 19th and start of the 20th century and what we do have are photographs showing the construction of the new GWR South Wales Direct main line from Wootton Bassett, just west of Swindon through to Stoke Gifford north of Bristol and running via Badminton and Chipping Sodbury.

Whilst some mechanical aids in the form of steam excavators and contractors steam engines were now available, there was still a considerable amount of manual labour involved - it was said that a farm labourer despite being recognised as being of 'stout physique' still took two years to reach the standard of fitness of a railway navvy.

These three images form part of a larger collection taken by a clergyman whose role it was to minister to the needs of the men and their families who often lived an itinerant existence in temporary accommodation as the work progressed. In the case of this particular construction, the contractor was Messrs S Pearson and the work took six years to complete. The views show construction of a cutting, early stages in excavating a tunnel mouth and work progressing within a tunnel complete with temporary narrow gauge track. *(David Foster Smith)*

It remains a matter of regret that no original broad gauge engine survives to represent the Great Western Railway in the years 1835 to 1892. There is an excellent replica at the Didcot Railway Centre. Two engines had in fact survived after 1892, *North Star* and *Lord of the Isles* but both were sadly dismantled in the early years of the 20th century for no other reason than lack of space at Swindon at a time when the works output was rapidly increasing. Even so not all was totally lost as some parts from *North Star* were retained and included in a non-working replica for the 1925 Stockton & Darlington celebrations. The replica also accompanied the brand new *King George V* on the latter's trip to America in 1927. A few years later it was one of the star exhibits in the GWR's own centenary celebrations of 1935. Once the pomp and circumstance had ceased the replica was retained and for many years stood on a plinth inside the works. For a short time in the 1950s it was on display outside of Swindon town hall before returning to the works. With the works closure in 1985 it was moved to the railway museum and now resides at STEAM Swindon. The 8' driving wheels are from the broad gauge engine *Lord of the Isles* and illustrate a time when it was considered the larger the wheel, the faster the engine might travel. These too were saved when the works closed and may be seen in the foyer at STEAM Swindon.

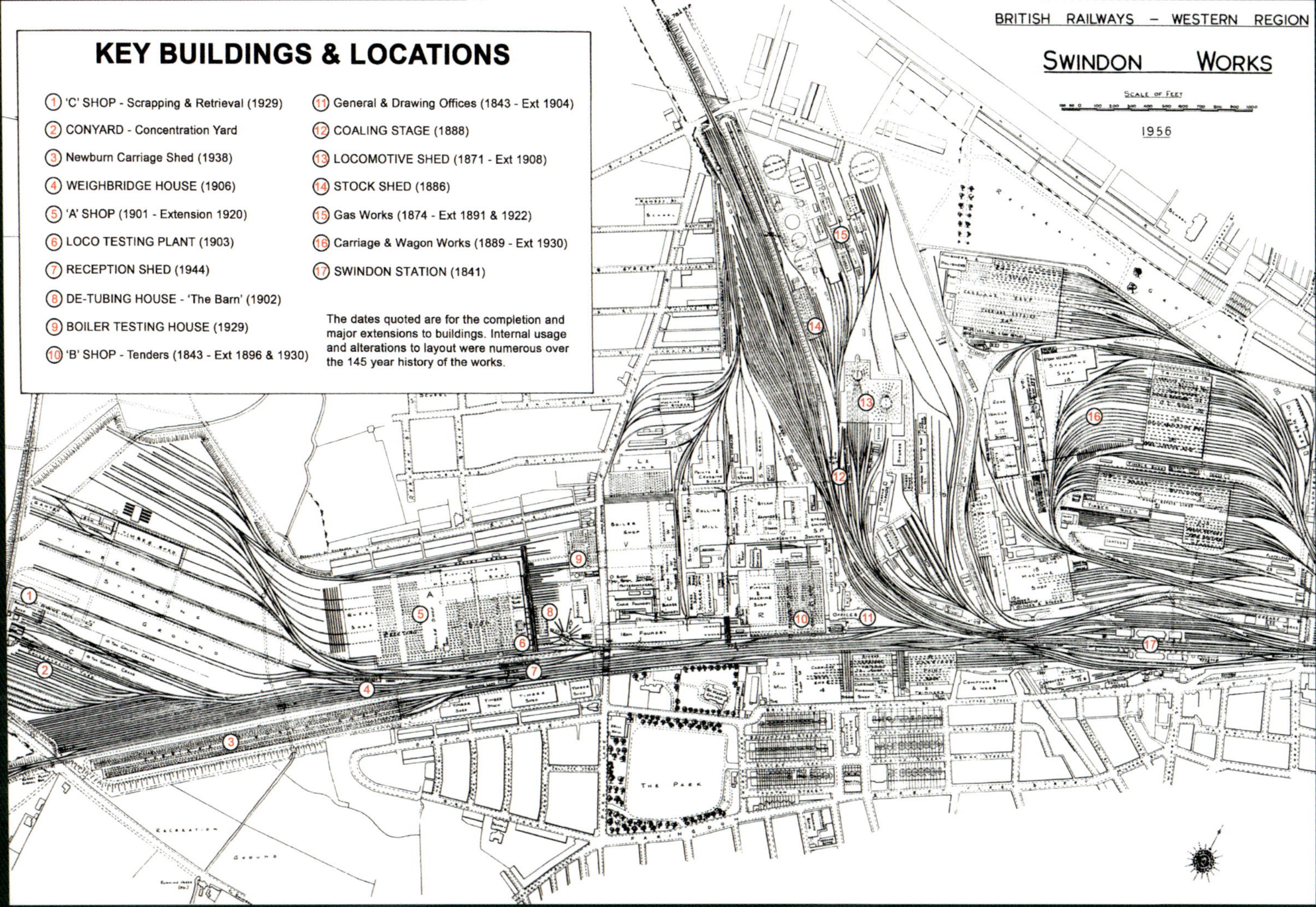

(Courtesy Andrew Malthouse)

Without doubt one of the most famous GWR engines has to be *City of Truro*. Built as No 3440 in 1903 it was renumbered 3717 in 1912. As No 3440 it is credited as being the first British steam engine to exceed 100mph having been recorded at 102.3mph descending Wellington bank in 1904. The speed was not made public at the time for fear of alarming passengers and only really came to light many years later. Since that time the record has been hotly debated - never of course in the minds of the true GWR enthusiast - but even if it was not 102.3 it was certainly very close to it. Such an achievement was enough to see the engine retained upon withdrawal in 1931 although not by the GWR and instead donated to the new railway museum establish by the LNER at York In 1957 it was called back from the museum and restored to running order at Swindon for service on normal and special duties. It is seen here at Swindon on 18 August 1957 soon after being returned to working order. *(L R Freeman)*

Not as famous but to the railway equally valuable. This is 'small Prarie' No 4507 at Swindon Works on a rainy November day, 15 November 1959. The 45xx class totalled 75 engines, built between 1906 and 1924 and were used mainly, although not exclusively, on branch line services. No 4507 is seen painted green and probably after an overhaul. It was built in May 1907 and for the last 15 years of its life was based in the West Country at Weymouth, Taunton and Yeovil. It would meet its end when withdrawn from service on 10 October 1963 and was scrapped in June of the following year. *(Transport Treasury)*

Left: A wonderful publicity photograph of seven 'King' class locomotives posed in numerical order 'ready for the summer service'. The engines are (L-R) Nos 6005, 6008, 6017, 6020, 6022, 6023 and 6024. From the line up, the last two *King Edward II* and *King Edward I* survive, both having been rescued from a South Wales scrapyard and restored to working order. A third member of the 30 strong class, No 6000 *King George V* is on display at STEAM Swindon.

Right: In later years No 6004 *King George III* after overhaul at Swindon and the fitting of a double chimney. This alteration allied to the not so obvious high-heat four-row superheater transformed the performance of the class in their final years until finally ousted by the diesel-hydraulics at the end of 1962. No 6004 is seen awaiting reattachment to its tender. Built in July 1927 it had a life of 34 years, 11 months and 20 days during which time it covered 1,917.258 miles. *(L R Freeman)*

The rather austere frontage of Swindon station as it appeared for more than 125 years, together with obvious signs of piecemeal extensions. Public access was on the down side and a subway with steps leading up to the platforms. The American car is definitely not representative of the day - the Morris Minor more so.

This time we are at platform level looking east towards Didcot, Reading and Paddington. MAS (multiple aspect signalling) has replaced the semaphore signals that once existed whilst at the same time 'rationalisation' has taken place with the original down platform out of use and all trains concentrated on the island to the left. This of course meant that down trains would have to cross and then re-cross the up line; in some respects a sop to Brunel's original single sided stations. In 1972 the down side buildings would be demolished; little more than wanton vandalism at a time when economy was more important than preserving heritage for future generations. The proverbial wheel would turn full circle again a few years later with the down side reinstated for passenger use allied to a modern station building. Unfortunately this also served as the ground floor of a large BR office block which still stands at the time of writing, 2025, but stands empty and cannot be used for its intended purpose due to the presence of asbestos.

The refreshment rooms at Swindon in 1929; the days when there was most definitely waitress service and items such as 'toasties' and 'wraps' had never been heard of. Here one partook of 'breakfast', a 'light luncheon', or afternoon tea, the latter no doubt including tea-cakes and biscuits made especially for the GWR by Huntley & Palmers. All served on plates containing the GWR crest, tea in cups and saucers and with silver ware similarly embellished. The British Rail sandwich was a nightmare of the future. In the earliest days of the GWR and for many decades after there was a compulsory 10 minute stop for all trains at Swindon and where passengers might partake of the delights (?) available. Mr Brunel was far from complimentary referring to coffee as tasting like roasted corn adding "I have long ceased to make complaints at Swindon. I avoid taking anything there when I can help it." Eventually the GWR bought out the rights previously enshrined with the refreshment rooms and which came about around the same time as a general speeding up of train services occurred. One surviving item from those earliest days is a coffee pot in the shape of a broad gauge engine and now in the care of the STEAM museum.

The 'Great Western Hotel' at Swindon located immediately opposite the station entrance. This is a listed structure with parts dating back to c.1870 and with a number of subsequent extensions. Folklore has it that when opened the pub was popular with the station porters as the nearby 'Queens Tap' were charging more at 2d per pint. A large wing of bedrooms was added in 1904, it remaining as a pub and hotel. Between 1974 and 2015 it bore two other names, 'The Noah's Ark' and the 'Flag and Whistle'. The original name was restored in 2015, although nowadays often shortened to the 'GW'.

If Swindon had become the spiritual heart of the Great Western then in 1881 that heart was due to be pierced by the thorn of an interloper, the Swindon, Marlborough & Andover Railway. This independent company proposed a line south and despite protestations from the GWR managed to open its route between Swindon and Marlborough in 1881.
The SMA station at Swindon was in what was known as Old Town and whilst there was a rail connection between the GWR station and SMA stations at Swindon the GWR were to say the least hostile to the newcomer fearing that when its line was complete it would loose revenue from important north-south traffic. Accordingly it set a charge of £3,900pa (the equivalent of £613,000+ in 2025) plus other costs for the SMA to use the GWR station at Swindon. Arbitration would considerably reduce this amount later.

Meanwhile south of Swindon the SMA had come to an abrupt halt at Marlborough and where the Board of Trade refused to allow through running over the Marlborough Railway until a number of improvements were made. These were certainly not hurried and it was not until 1883 that a through service was finally possible from Swindon to Southampton, running over the GWR from Marlborough to Savernake and then the London & South Western Railway from Andover to Southampton - the SMA having more cordial relations with the LSWR.

Relations with the GWR however remained poor not helped when a further company, the Swindon and Cheltenham Extension Railway came on to the scene with a line from the SMA at Swindon north to Andoversford which would allow trains to run direct from Cheltenham to Southampton. In connection with the northwards extension the company name was also changed to that of the Midland & South Western Railway.

Despite though now having a through route, difficulties with the GWR persisted and an approach to the LSWR for assistance secured the services of an extremely competent new manager (Sir) Sam Fay. Fay quickly took stock of the situation and whilst able to achieve much the difficulties imposed by operating trains over the GWR between Marlborough and Savernake were considered to be the prime obstacle. In consequence a new independent line was built between Marlborough and Grafton south of the GWR line at Savernake. This included a deep cutting and tunnel and was opened in 1898. Fay was also able to negotiate on friendlier terms with the GWR. It also resulted in both Marlborough and Savernake having two stations.

Swindon too had two stations, that on the MSWJ line becoming known as 'Town' the GWR station eventually becoming 'Junction'. The MSWJ route was subsequently absorbed by the GWR in 1923 and survived intact until 1961. Short sections only then remained to the former power station at Moredon, to Cirencester and to the Town station - the latter's last use as a railhead for stone during construction of the M4. In the Swindon area all of this has now gone and much of the route disappeared under development. North of Swindon part does remain as the heritage, 'Swindon and Cricklade Railway' whilst further south the military operate a short stub from Andover as far as Ludgershall.

Duke class 4-4-0 No 3285 *Katerfelto* leaving a smokescreen as it enters Swindon Town station from the north bound for Marlborough, Andover and Southampton probably c.1936. The light load consists of just two coaches and a ro-rail wagon on which is loaded a milk tanker. This is probably destined for Andover and will then be transferred to the Southern Railway continuing its journey to one of the London creameries. This was a typical train over the former MSWJ line in the 1920s and 1930s, two-three coaches and perhaps a horse box or milk tanker. In addition to the through trains there was also a good shuttle service between Swindon, Marlborough and Savernake - for connections to main line services. The building on the right is the former headquarters offices of the MSWJ and survives today. As an independent company the MSWJ operated its own locomotives and stock, much of the latter originating on the Midland Railway. Infrastructure requirements, including the two signal boxes at Swindon and also that at Rushey Platt were supplied by the LSWR. Subsequent to the takeover by the GWR the route was slowly 'westernised' hence the GWR signals seen at the north end of the station. Sadly although three of the former MSWJ locomotives managed to survive until the early 1950s their historic importance was not identified and these relics were lost to scrap. No 3285 similarly did not fare well, built by the GWR as No 3219 in 1899, the number change occurred in 1912, but it too had been withdrawn in 1937. *(Milepost)*

The final day for the former MSWJ station at Swindon Town. No 6395 has charge of the last through train from Cheltenham St James to Southampton. A similar working traversed the line in the opposite direction. The date was the 9 September 1961.

Derelict and abandoned. Never again will the directors of the MSWJ look down on their railway. This is the remains of Swindon Town in the period after stone trains stopped in 1972 and the rails lifted in 1978. Where once porters carted luggage and passengers started their journeys north, south and even father afield nothing remains. The MSWJ exists now only in photographs and memories.

Left: Back to the GWR proper and the Works itself sometime in the 1950s. The engines are in for various work, then known as 'heavy', 'heavy intermediate' or 'light' repairs. That in the centre is a BR Standard type; identified by the reversing wheel on the left indicating left-hand drive. (Great Western engines without exception were right-hand drive.) Left and right are indeed two GW types. That to the left cannot be identified due to the supporting roof pillar, right may be either a 'Hall' or 'Manor'. Note the steps and cages used to allow fitters access to the footplate. This was how Swindon was for over a century. *(Transport Treasury)*

Above: No 7924 *Thornycroft Hall* being attended to again sometime in the 1950s. Built under the auspices of British Railways in 1950 this engine, like so many others from this era, was destined to have an unduly short life of just 15 years. For the final months of its life it was based at Oxford after which and again like several of its classmates, it was towed to Messrs Cashmore at Newport and broken up. *(Transport Treasury)*

Below: Ian Allan Locospotters Club Special to Swindon 1 September 1949. In the grey days post WW2 as Britain slowly recovered from conflict, a day out was a welcome distraction. Starting from Paddington several hundred booked for the journey to Swindon and a tour of the Works. In case it was needed there was also the added incentive of a chocolate biscuit - chocolate then being on ration. Here No 6021 *King Richard II* emblazoned with 'British Railways' on the tender has arrived with the special and the passengers are gathering around to watch what will shortly be an interview by the press with Mr Allan. Nowadays enthusiast tours and special trains are commonplace but in 1949 this was a rarity indeed. *(Transport Treasury)*

Right: Within the Works complex was the Locomotive Testing Plant. Instigated in the time of Churchward this facility could see a huge engine appearing to move at speed and yet stationary; the wheels turning rollers and from which in turn could be measured power output gauged against coal and water consumption. (The tender has been removed and instead a 'driver' is working in the cab whilst the engine is connected to a measured water supply and weighed coal is fired as needed.) Here No 5018 *St Mawes Castle* is working flat out and being observed by a number of staff, none of whom are far away from the whirling rods. *St Mawes Castle* was built in 1932 and moved around the GWR and WR seeing service literally north, south, east and west until ending its days at Reading in March 1964. It would not survive into preservation. *(L R Freeman)*

ST MAWES CASTLE
5018

Left: One of the more unusual scenes at Swindon was the occasional visit of one of three Vale of Rheidol steam locomotives for repair. The Vale of Rheidol line is a narrow gauge, 1' 11¾" railway running for 11¾ miles from Aberystwyth to Devils Bridge which opened in 1902. Built originally to carry lead ore this traffic was declining even before the railway opened and instead it quickly found a new type of traffic - passenger tourism. Running a railway is not an easy undertaking and the original company sold out to the Cambrian Railways in 1913 and which in turn became part of the GWR in 1923. In 1948 British Railways were in charge and which continued to operate the line using the three steam engines that had been at work for many years. After 1968 BR would market the line as being operated by British Railway's only steam engines, the trio painted in corporate blue with the double arrow on the tank side - not a good combination. The VoR was the first BR privatisation, being sold in 1989 and has flourished once again under private ownership. Seen here is No 7 *Owain Glyndŵr* in BR green livery on a Flatrol wagon post overhaul awaiting return to Wales. *(L R Freeman)*

Right: Crane tank No 1299 at Swindon. Crane tanks were used for lifting loads in connection with engineering purposes, the GWR having four. No 1299 originated on the South Devon Railway but was in an incomplete state when that company was absorbed by the GWR. No 1299 was fitted with a crane in 1881 continued at work until 1936. *(R C Riley)*

1299
GREAT WESTERN

L.95
LONDON TRANSPORT

Left: London Transport pannier tank No L95 one of 13 similar engines bought by London Transport from British Railways between 1956 and 1963. LT had always run some steam engines operating works and engineering trains although by the mid 1950's their own fleet was starting to show its age and in need of replacement. In consequence trials with the Swindon design took place in 1956 and which subject to modifications being made to the cab and other fittings were successful. Once details had been sorted out several more were built taking the numbers L89 through to L99. The original Nos L90 and L91 were subsequently substituted the replacements taking the same numbers. The engines were based at Lillie Bridge and Fulham the last three active until 1971. By this time it had been some years since steam engines had been used by British Railways in London; Waterloo the last steam terminus back in 1967. An LT pannier tank was thus responsible for a letter to one of the daily newspapers, the correspondent commenting he had been woken by what he was certain was a steam engine whistle but yet was convinced no steam engines had worked in London for many years .... Such was the success of the pannier tank design that during BR days examples might be seen away from their usual haunts at Waterloo and as far afield as Folkestone, and even working trains on a remote branch line in Scotland. A number of WR pannier tanks were also sold out of service for use by the National Coal Board. The pannier tank design had originated on the GWR back in the early years of the 20th century when Belpaire or square top fireboxes rendered the use of the previous saddle tank design incompatible. Eventually the design was used on numerous classes with well over 1,000 pannier tank engines built or converted for the GWR. No L95 was originally GWR/BR No 5764 having been built in 1929 and sold to LT in 1960. After withdrawal in 1971 it was sold to the Severn Valley Railway. Note when the photograph was taken at Swindon No L95 appears to be the only steam engine present, the works otherwise entirely concentrating on diesel types. Notwithstanding having originated from Swindon, London Transport pannier tanks were also repaired at Eastleigh when required. *(Transport Treasury)*

Right: The cab of No 6014 *King Henry VII*, unique for a member of the King class in having a wedge shaped front, part of a number of 'streamline' changes - one can hardly call them 'improvements' - added in March 1935. 'Castle' class No 5005 Manobier Castle was similarly altered. No 6014 had been modified with a form of streamlining it is said at the behest of the directors who had heard that the London & North Eastern Railway were about to introduce their own streamlined engine - the 'A4' class. Accordingly Mr Collett at Swindon was instructed to respond, and rumour has it he did so with a lump of plasticine on a model of one of his engines. The results did little for the engines and even less in the context of true streamlining. It was not long before the various parts were removed although No 6014 retained the wedge shaped cab - note too the hand-operated wiper - for the remainder of its existence. *(Transport Treasury)*

Below: Apart from locomotives Swindon was also responsible for a colossal output of carriages, these commencing as small four-wheel vehicles, then on to six wheels which rode better, and as design progressed to bogie vehicles. As the years passed so different types also emerged; individual class coaches, composites, restaurant, sleeping and 'special duty' types - see page 30. Originally too the basic passenger carrying coaches of the 'all compartment' type, the corridor coming later as did through connections between trains. Compartment type vehicles with a side corridor would remain the norm well into the 1950s. As more modern stock was built so older vehicles might be cascaded on to branch line use or scrapped but for some there was a reprieve with the GWR Camp Coach scheme where a specially fitted-out coach having bunk beds, a communal dining room and kitchen, would be assigned to a particular rural location for the summer season and which could be hired out by a family for their annual holiday. The choice of places was most definitely rural the only interruption to the sound of bird song being the occasional branch train.

Camping in the 1930s was very different to 'glamping' today with no running water, cooking on a solid fuel stove and no heating. Even so they proved incredibly popular and up to 1939 bookings thrived. The idea was resurrected in the 1950s including having several larger vehicles, as above, stationed on a single site at Dawlish. (The vehicles returned to Swindon during the Winter period for maintenance.) In the view below we have a former 'Toplight' dating from the first decades of the 20th century freshly overhauled and repainted and ready to return to its allocated location.

Opposite: Peace and tranquillity at Luxulyan in 1960. In some respects it is a pity some of the heritage lines have not revived the idea albeit with mod-cons. Somehow though I cannot see it catching on at a former station on the main line network with trains passing at 125mph every few minutes. *(Transport Treasury and R C Riley)*

CAMPING COACH

At rest in the yards at Swindon after a lifetime of service is W7W, a unique vehicle and the GWR's original Dynamometer Car. This coach had been built in 1901 and was used for on-road locomotive testing, being able to accurately calculate drawbar horsepower and with the aid of a number of attachments to the locomotive, replicate steam temperatures, vacuum etc. In this way locomotive efficiency could be calculated and the results of modifications assessed. It was designed with a corridor connection only at the one end and also had what is referred to as a 'Royal Clerestory' meaning the centre section of the roof sloped down at the ends. It reached the end of its life in the early 1960s and was replaced by a modified Hawksworth vehicle. Minus its specialist testing equipment it now survives on the South Devon Railway. *(John Tolson)*

The GWR were the pioneers of successful diesel railcar operation in the 1930s, No 4 seen below, an example of the streamlined type. These vehicles were built in connection with Messrs AEC at their Park Royal workshops and in service proved very popular on a number of services; indeed so much so that at times a locomotive hauled service had to be substituted. Over the years the design morphed to a razor edge type and eventually to twin sets into which a centre coach could be added. Three cars survive, W4, seen here restored to GWR livery as part of the National Collection, No 22 at Didcot, and No 20 on the Kent & East Sussex Railway.

W9553W is a 70' diner from Lot 1177, Diagram H19 of 1911. Four vehicles were constructed to this Lot and in company with some similar length First-Class vehicles and Parcels vans were designated as 'Fishguard Boat Stock'. The coach is seen here at Newton Abbot on 19 July 1957 in company with a Hawksworth Brake and BR Mk1. This particular coach was reconditioned in 1936 and as well as new windows (the rounded corners were not dissimilar to the style later adopted by Bulleid) replacement 9' plate-frame bogies were fitted and side panelling applied. It is believed the other vehicles from this Lot were similarly modified around the same time, although it will be noted the retention of ventilators above the door droplights presents a somewhat dated appearance. Weighing 40 tons, this vehicle was withdrawn in 1958. *(R C Riley)*

Former Restaurant Car No. 9527 recorded as departmental No. 079128 (both the 'W' prefix and suffix are superfluous) outside Swindon shed, 10 July 1963. This vehicle had been completed on 23 June 1906, one of a pair of Concertina type First Diners to Diagram H14, Lot No. 1115. At the time, these massive 70' x 9' vehicles - hence the recessed doors, were carried on 4-wheel bogies and took the running numbers 405/6, shortly afterwards changed to 8405/6 and then post 1907, to 9526/7. (The 'W' letter came after 1948.) The car was rebuilt and modernised around January 1939 at which time it also acquired the 6-wheel bogies seen here. It was still in use operating between Swindon and Gloucester circa 1960 but found its last use seemingly as office accommodation; from outside at least it seems to have retained some of its interior fittings including table lamps. At one end is 'Western' class diesel-hydraulic No D1054 *Western Governor* built at Crewe in 1963 and withdrawn after an unnecessarily short life in 1976.

The eight 'Super Saloons' built by the GWR in 1931 rank amongst the best known and most easily identified of all the Swindon built vehicles. The type had its origins in the 1929 'Riviera' stock, whilst the later 1935 'Centenary' coaches again displayed a family likeness. Their purpose was a simple desire by the GWR to afford a luxury service at a cheaper rate than had been achieved with the hire of Pullman vehicles and attendants, as tried out on the 'Torquay Pullman' during 1929. In addition Pullman coaches had been included in the formation of some of the Plymouth boat trains. With their arrival in service, the new saloons took over the prestige workings and although officially known as 'Super Saloons' the names 'Pullman', 'Cunard' or 'Ocean Saloon' stock was used. The vehicles were also much used for private-hire and on the Member's Newbury Race trains. Although rarely reported as running as a rake of eight vehicles plus brake(s) (no brake-end super-saloons were constructed), it was not unknown for one or more to be included in the formation of a 'Royal', especially when this was included as part of an ordinary service train. Their use on Plymouth workings continued until as late as the end of 1962, after which BR were no longer able to deal with liners at Plymouth. They were then to be seen employed for party travel, as well as continuing on the Newbury workings referred to. Possibly running for a special party, W9117W, the former *Princess Mary*, is seen at Snow Hill. Under the GWR, all eight vehicles had been named, although these ceased to be carried sometime after 1948. Post 1948 under BR, the livery was changed to standard 'crimson and cream' although all were repainted in WR chocolate and cream from 1957 Fortunately five of the eight have been saved for preservation. *(Amyas Crump)*

In 1958 three Hawksworth Brake-Composite vehicles to the same diagram (E164) and Lot as that seen left, were converted to double ended slip coaches with the corridor connections removed, the requisite observation windows added and fitted with additional vacuum cylinders. These were Nos 7374-6 which retained their original numbers. By this time only three trains had slip portions, the Weymouth slip off the down 'Cornish Riviera' - slipped at Heywood Road outside Westbury, the up 7.00 am Weston-Super-Mare to Paddington which slipped at Didcot, and the 5.10 pm Paddington - Wolverhampton which slipped at Bicester. All three vehicles were painted in WR chocolate and cream although at least one (and probably the others), No. W7376W had appeared in standard maroon by 1963. Slip coach working ended with the Bicester slip on 9 September 1960 after which all three had their slip couplings replaced with standard draw-gear and were transferred to Taunton in early 1961 for use on the secondary / branch lines to Barnstaple, Minehead, and Chard as well as on main-line stopping services. In the illustration above, taken at Barnstaple Junction, and after a return to conventional duty, the additional vacuum cylinders will be noted. At left is an example of a Corridor Third, having eight compartments and built to Diagram. C77 around 1938. *(Brian J Harding)*

Left: As mentioned, the GWR also built thousands of goods vehicles, the movement of goods being the principal source of revenue for the railways for many years. Aside from their individual running numbers, goods vehicles were identified by a telegraphic code word, this example being a 'Mink F', a closed van carried on two bogies. Seven of this type were built in the early part of the 20th century but the design was not perpetuated further. *(Transport Treasury)*

Below: One of the oddities to be found at Swindon was No 26, a 4-wheel petrol shunter although not unique as there were others including one that operated at Taunton. Obviously because of their size they were limited in haulage ability but came in useful in moving light loads when required. Photographed on 16 June 1957, it was still extant in the early 1960s although likely unused for some time. *(R C Riley)*

6300
FREDY
DONE
THIS
YOGI

Left: 63xx No 6300 at Swindon having had a 'coming together' - we are not told where, how, and when nor how the other party fared. The engine has arrived at Swindon for assessment which was not good news as it was condemned on 27 June 1960. It is seen here outside the works four weeks later on 24 July 1960 pending scrap. According to the chalked inscription on the cylinder, evidently 'Fredy' was responsible. *(Transport Treasury)*

Above: A happier visit to Swindon for one of the big 72xx 2-8-2T engines, No 7245. Originally built as a 2-8-0T in 1912, it was taken into Swindon in 1938 with an extra pair of wheels added at the rear to support a larger coal bunker and additional water capacity and as such could work over an extended range. Although most were based in South Wales a few could be found elsewhere, No 7245 being one of the South Wales engines and which survived until October 1964. *(Transport Treasury)*

From the largest tank engine to one of the smallest, No 1338 originally built for the Cardiff Railway in 1898 and subsequently passing into the hands of the GWR and then BR. It is seen here fresh from overhaul complete with the first BR emblem - even if it does look a little as if it is perched on the tank side. After a life of 66 years No 1338 was saved from scrap and moved to a private museum on the platform of the former Bleadon & Uphill station. Vegetation was on the point of swallowing it up when it was moved to Didcot in 1987 and restored to working order although for the present the boiler certificate has expired. *(Transport Treasury)*

A happier outcome for a Mogul compared with No 6300 seen previously. Here No 7340 has been overhauled and received smart green livery as was the case in the late 1950s. It also carries the later (post 1957) BR emblem on the tender. This engine had originally been built in April 1922 and given the number 9318. The engines in the 93xx series had extended frames with extra weight on the front end so meaning they were placed in the 'Red' route category. (GWR engines were given colours from 'Double Red' to 'Red', 'Blue', 'Yellow', and 'Uncoloured' to indicate their route availability.) No 9318 was rebuilt with reduced weight and renumbered as seen on 31 December 1957, after which it was sent to Wales where it wandered around the depots at Neyland, Llanelli, and Carmarthen before ending its days at Didcot on 15 June 1964. *(Transport Treasury)*

Another engine fresh from shops, this time No 6011 *King James I* and from the '84A' shed plate displayed ready to return to Plymouth Laira. The 30 'King' class locomotives were employed on the heaviest services until finally ousted by diesels in 1962. Even so it was very much a case of one department not talking to another as engines were being called into works for overhaul only to be withdrawn just months later and potentially with considerable life left; but such was the pace of change. Indeed it was around this time that a memorandum went out to state that if one of the new diesels were to fail - a regular occurrence at the time - then the public announcement should only refer to a 'locomotive' and not a diesel failure. No 6011 and No 6934 *Beachamwell Hall* both stand out compared with others nearby; and of course one wonders what else was behind and in the distance...? *King James I* entered service in June 1928 and was one of the last of the class in service, being withdrawn from Old Oak Common on 18 December 1962. It was scrapped at the place of its birth, Swindon. *(Transport Treasury)*

Away from clean engines and to a workaday example at Swindon station. This is 38xx class No 3827 in charge of an express freight which includes a number of 'fitted' vehicles; meaning these were braked and the brake controlled from the engine. Indeed the first part of the train looks very much as if the majority are cattle or livestock vans. The railways were regular hauliers of livestock with regular stops made for watering and other requirements. Swindon was a place where engine crews were often changed although it is unlikely this was the case here as the ATC horn would be sounding as the engine is passing/standing on the ramp - just seen between the rails - and the distant signal is 'on'. Had it been a crew change then the engine would also be likely to be taking water. On the platform line is a pit where drivers could oil and examination their engine whilst stopped. *(Transport Treasury)*

Swindon was responsible for building the 45 engines in the BR standard 82xxx series, the design basically an amalgam of an LMS chassis and Swindon boiler. Intended for branch line work the class had a short life their duties either lost through closures or superseded by diesel traction. None survived although a 'new-build' is making progress at Bridgnorth (2025). Alongside is pannier No 5737 and with a definite 'Not to be Moved' board predominantly displayed. 18 August 1957. *(Transport Treasury)*

Former Taff Vale 0-6-2T No 397 at Swindon on 18 August 1957 and probably in use as a works shunter; certainly the express lamp code is not applicable! Alongside is 57xx 0-6-0T No 5741 withdrawn from service two months earlier and likely performing a similar function. *(Transport Treasury)*

Apart from the works, there was also a sizeable steam shed at Swindon and it was here that ROD (the name for the former WW1 Railway Operating Department) 2-8-0 No 3028 was photographed alonsgide what had been the oil fueling area. Nop 3028 was one of a number of these engine purchased by the GWR after WW1, some literally run into the ground and scrapped, others retained for freight use. It has to be said they were not universally popualr with their crews, a certain bias against what was not a Swindon product perhaps but also because they would leak steam in the cab and seemed to have just one speed - a plod. Even so they could haul prodigious loads with many, including No 3028, lasting well inthe 1950s. *(Transport Treasury)*

Engines at the end of their life in 'C' shop also known as the 'con' (for condemned) yard. Hard to realise that today's car park for the Outlet Centre was once laid with sidings and where the broad gauge engines and stock of the 19th century were concentrated after 1892. Locomotives that had reached the end of their life were shunted here after having been assessed as being of no further use. Back in the 1930s a number of these might be placed on what was known as the 'Sales List' in the hope of finding a buyer; there were few, but oh for the ability to turn the clock back! Here three engines at least have reached the end of their days, former 'Dukedog' No 9022, 53xx No 5314 and which still retains at least one cabside numberplate, and behind another 'Dukedog' 4-4-0. Boilers, which were standard parts amongst several classes, were sometimes saved for further use. At one time Swindon dealt with all of its engines due for scrap but as the pace of withdrawals increased so engines were sold to various metal dealers the most famous being the Woodham's yard at Barry where rows of engines slowly rusticated from the effects of salt air and that of the souvenier hunter. Woodham's would provide a rich source for the preservationist with a number of former GWR engines rescued, restored and running today. *(Transport Treasury)*

Above: With sparks flying this is the end for 0-6-0T No 5419 on 15 March 1958. Entering service in June 1932 it ceased to work from Westbury in February 1958 and as seen here did not have to wait long for its destruction. *(R C Riley)*

Right: Footplate of a GWR 4-6-0 and with the chalked mark 'arch in' meaning the brick arch has been installed in the firebox. The cabs and controls of all GWR engines were similar although it must be admitted, crew comforts were limited. *(Transport Treasury)*

Mention was made earlier of the absolute waste of resources that occurred at times due to a lack of co-ordination between departments. Here we see No 6025 *King Henry III* fresh from overhaul; engines would leave the works area on traverser, a corner of the rails from which are just visible in the bottom left corner. They would then be moved to a steaming bay and reunited with their tender which had come from a separate shop. This view was taken on 4 March 1962, No 6025 spick and span and ready for further years work - yet it was withdrawn nine months later in December. *(Roger Holmes)*

2-8-0T No 5241 also fresh from overhaul. This was the form in which No 7245 had first appeared, see page 39. These big tank engines along with the smaller 56xx/66xx type 0-6-2T's were used on the heavy coal trains of South Wales although one or two were stationed in Cornwall to deal with chine clay workings. Built in 1924 it had a life of just over 30 years, ending its days at Newport (Ebbw Junction). Following withdrawal No 5241 did not make its way to Swindon but instead was broken up by Messrs Cohens at Morriston. *(Roger Holmes)*

The equivalent heavy freight engine but for longer hauls, No 3800 and perhaps a bit cleaner than sister engine No 3827 - see page 43. No 3800 appears to be having attention to the left hand piston valve perhaps after a trial run. Notice in the background the stored boiler; engine parts would be stored all over the site, boilers, wheelsets, chimneys etc. all to be called upon when required. *(Roger Holmes)*

6025

Left: No 6025 and behind No 6018 both mantled by snow on a bleak January day in 1963. No 6025 *King Henry III* we met earlier (page 49), No 6018 *King Henry VI* similarly taken out of service in December 1962. There had been thought of maintaining a reduced fleet of engines of the class to cover failures of diesels in the Summer of 1963 but this did not happen. In the event No 6018 did run again, but finally on 28 April 1963 on a railtour organised by the Stephenson Locomotive Society. Various attempts were made to preserve the engine later but these came to nothing and No 6018 was broken up at Swindon six months later. *(Roger Holmes)*

Right Top & Bottom: As mentioned, at any works there will be the unusual strewn around, this included this 'Denparts' wagon (Diesel Engine Parts) and which may have commenced life as a 'Con-Flat' (container flat wagon). It will be noted to be vacuum fitted so could at least move between Swindon and the various depots at a faster speed than in an unfitted service. Equally interesting is this tender from a former MSWJ engine - see pages 16 to 19 reference the MSWJ railway The last three former MSWJ engines in service were Nos 1332/3/4 at Reading in the early 1950s. This would appear to be the tender from one of them which survived as a water tank in the Reading area. The point of interest is how the original MSWJ scroll has survived through the paintwork and dominates the later 'Great Western' that had been applied on top. The MSWJ had its locomotive and carriage workshops at Cirencester and which were similarly taken over by the Great Western in 1923. For a time locomotives continued to be repaired there, the story going that post 1923 when an engine had been repaired the former MSWJ foreman put an order in to Swindon for copious quantities of green paint as the last item needed. Back came his allocation, a much smaller amount. The harassed foreman threw his arms in the air exclaiming, '...that is nowhere near enough, the heat will make the green turn to black as soon as it gets hot.' Apparently it did and Swindon in future had to increase their allocation.
(Transport Treasury and Roger Holmes)

18000
18000
PLATE

Post WW2 and prior to nationalisation in 1948 three of the four railway companies had aims towards a replacement for steam. So far as the GWR was concerned this was in the form of gas-turbine with electric propulsion. Two prototype machines were ordered, one from Metropolitan Vickers in Manchester and the other from the Swiss company Brown Boveri. The Swiss machine arrived first but not until BR days in 1950. It burnt heavy oil and was perhaps cruelly nick-named *Kerosene Castle*. A side effect of the fumes was said to be a certain laxative effect for passengers in the hauled trian. As a one off it did most of what was expected of it although it suffered from limited life to the combustion chamber and poor maintenance as it had to be serviced amidst the dirt of a steam shed. Even so it managed to survive for ten years after which it was presented to the International Union of Railways in Europe for use as a test bed on wheel adhesion - the gas turbine power unit having been removed. Having served its purpose it was moved to Vienna and put on display. The shell was subsequently repatriated and after several moves now resides at Didcot. In the view on this page, No 18000 is seen stored outside the Swindon stock shed. Below the same engine is depicted passing the Works with a Bristol service, a commonplace duty in its later years.
(John Bailey and Transport Treasury)

No 18100 had been the first of the two gas turbines to be ordered but was not delivered until 1951 and designed to burn ordinary fuel oil. It is seen here in the original black and silver livery - as had also been applied to No 18000. This engine did not have a nickname, although perhaps 'gas-guzzler' might have been appropriate as its consumption of conventional diesel fuel was excessive. When operating on full power its haulage capacity was prodigious; on test it started a 20 coach train from stationary on the steepest of the South Devon banks, but such power was rarely required especially as it operated on existing steam schedules. In service it too suffered the difficulties of being serviced in a steam shed whilst bogie cracks showed up soon after delivery. Another issue affecting both gas turbines was that due to rotas, drivers only came across the engines something like one week in 16 and so lack of familiarity issues were the cause of some failures. With little work it could usually perform without consuming excess fuel, No 18100 was withdrawn in late 1952 and returned to Metropolitan Vickers with the intention of converting it to also burn heavy fuel. This was never achieved and after several years the conversion was abandoned and instead it had a radical transformation into a straight electric locomotive for testing of the LMR 25kv electrification. It is depicted here in its original gas turbine guise in the works yard on 30 March 1952. *(L R Freeman)*

Modern (first-generation) diesel mechanical unit built at Swindon. These units were the replacement for steam on many local services and continued to operate well into the 1980s and beyond. The 'whiskers' on the front were intended to act as a warning of their approach but were later replaced by a yellow panel and then a full yellow end. *(Transport Treasury)*

End of the road for small Prarie No 4569 at Swindon in September 1964. Built in 1924 it is seen here stored with the Works Gas Works in the background. Already the cabside and smokebox numberplates have been removed, as also have the connecting rods. No 4564 would not last much longer and having worked for 40 years was scrapped in November 1964. *(Transport Treasury)*

Another engine to reach the end of the road in 1964 was large prarie No 6144 and clearly showing its blue route availably disc. These engines had been used on suburban services around London and Birmingham for almost 30 years but were replaced by diesel units of the type seen previously. With no further work withdrawal was inevitable. In happier times and clearly fresh from overhaul, No 6167 is depicted facing the front of Swindon running shed. *(Graham Smith)*

Below: No 6008 *King James II* stored pending scrap and yet retaining name and number plates - what price today? Mention was made earlier of how some components might be removed for further use and on No 6008 the speedometer drive has been removed from the rear offside wheel. Notice too the cover of the rear sandbox is hanging down and yet despite this apparent lack of care the engine is correctly stabled in mid-gear. A sad end to a proud machine.

Right: Yet it was not all doom and gloom. 'Dean Goods' No 2516 was withdrawn from Oswestry in May 1956 and a new regime compared with decades earlier had decided that it should be saved as a representative example of the class. It is seen here at Swindon in most definitely work-weary condition and having also been exposed to the elements for some time. In due course it was taken 'inside' to be lovingly restored and after a spell on display at the first Swindon Museum is now one of the exhibits at STEAM.

2516

Left: In previous images we have seen what was going on outside and so to redress the balance here are two views showing the inside of the cavernous 'A' shop with engines being built and also under repair. There is a new BR Class 9F 2-10-0 being erected, the number 92178 chalked on the cylinder. There were 251 of these massive of engines built, the majority at Crewe although Swindon was responsible for 53. Although intended for freight they were so successful that a number found use on express passenger trains, that is until contrary instructions were issued from BR headquarters. No 92178 was one of batch of six built in 1957 for the Eastern Region, it would have a disgustingly short life of just eight years. Swindon was also responsible for classmate No 92220 in 1960, the very last steam engine to be built at Swindon and indeed for British Railways; it was appropriately named *Evening Star*. No 92220's working life was even shorter at just five years although it was retained for preservation and in 2025 was on display at the National Railway Museum. A new 'Warship' diesel hydraulic can be glimpsed under construction to the right. *(L R Freeman)*

Right: In Great Western days when steam was in its heyday in 'A' shop. Engines of all shapes and sizes in various stages of repair. Those identified by number include Nos 4591, 302?, and 6370.

56
57
4591

Below: Train spotters take engine numbers - would that extend to smokebox types? This is another of the outside dumps at Swindon, the different diameter smokeboxes corresponding with different boiler types. The GWR had been masters of standardisation for years; the standard No 1 boiler for example used on the 28xx, 29xx ('Saint' class), 40xx ('Star' class), 49xx ('Hall' class), 68xx ('Grange' class), and 6959 ('Modified Hall') types. Hard to imagine that where industry once thrived this area is now roads and housing. *(Roger Holmes)*

Right: 'Metal Millinery' of various types (and sizes) although more accurately dome and safety valve covers of differing types. It was known as the 'Swindon 'Hat Rack'. *(Transport Treasury)*

Left Top: We come now to our last two pages of steam and a sad sight at Swindon on 24 June 1962, a line up of seven condemned 'Castle' class engines, plus a solitary 43xx/53xx type. Only one can be identified by number, No 5024 *Carew Castle* and officially withdrawn the previous month. (The name is that of a castle and tidal mill in Pembrokeshire.) No 5024 would be hauled away to South Wales and broken up in December 1962. At this time some trains were still of course steam operated and it used to grieve crews battling away on an engine that was clearly not in the best condition to see better members of the same class lined up on the scrap roads. *(R C Riley)*

Left Bottom: A line up of former Eastern Region 'V2' class locomotives at Swindon for scrapping - WR types behind. These engines had arrived in Wiltshire to give up their tenders which were to be converted into propelled snow-ploughs. With the tenders removed the engines were of no use and they met their end here far from their native Doncaster. Although one is identified, No 60941, it was not the first time an engine of the class had been to Swindon. No 60845 visited in 1953 for performance and efficiency tests which saw it running on the test plant as well as hauling loads of up to 25 coaches (762 tons) between Reading and Stoke Gifford. This was made possible as a result of some relatively small modifications to the blast pipe and chimney dimensions which more than doubled the rate of continuous evaporation that could be achieved. *(Transport Treasury)*

Right: Whilst on the opposite page we saw a line up of withdrawn 'Castle' class engines, here is one that escaped the cull very early on. No 4073 Caerphilly Castle seen here on display in the STEAM museum at Swindon complete with original small tender and the 'Chetenham Flyer' headboard. *(Andrew Royle)*

CHELTENHAM
GWR
FLYER
WORLD'S FASTEST TRAIN
4-6-0

A change to 'modern' traction. 'Western' class diesel No D1002 *Western Explorer* alongside 'Castle' class No 5091 *Cleeve Abbey*, standing outside the running shed at Swindon on 3 June 1962. As will be seen opposite, the 'Western' class engines appeared in several liveries over their lives but this and the alternative maroon were probably the favourites. D1002 entered service from Swindon on 19 March 1962 and was the third of the class. In 1965 it was repainted maroon and then corporate blue in 1970. It survived until 1974 and was broken up at the place of its birth in the same year. No 5091 was slightly longer lived, starting off as a 'Star' class engine, No 4071 in February 1923 but converted to a 'Castle' at the end of 1938. It had a 42 year life and was withdrawn at the end of 1964 but cut-up at Messrs Cashmore's at Great Bridge. *(Transport Treasury)*

First of its class, No D1000 *Western Enterprise* at Swindon sometime between the end of 1962 and October 1964. The 'Western' class were one of the designs that had taken advantage of the BR Design Panel, an organisation within the railway intended to ensure new products basically were pleasing to the eye. Matters had come to a head with the 'Metrovick' design a few years earlier which were the total opposite when it came to appeal - but then beauty is in the eye of the beholder of course. D1000 is wearing its 'desert sand' livery, the Western Region as ever adopting an independent streak. That would all change when Stanley Raymond came to Paddington in 1962/63, determined to take the independence out of the (Great) Western Region; post 1964 - corporate maroon all round. *(Transport Treasury)*

An example of an attractive locomotive design from the 1960s. The Beyer-Peacock 'Hymek' class of which 101 were built between 1961 and 1964. D7003 is one of the earliest engines dating from 1961 and is seen with a green liveried D800 series 'Warship' behind. The two tone also featured on the diesel-electric D15xx (Class 47) loco type. Unfortunately one feature that did not stay clean were the bogies. A combination of brake dust and deposited toilet material (at a time when all trains had toilets with 'bottom' discharge, meant a murky brown became more common. As originally built original Western Region (contact) ATC was fitted. After an eleven year life '7003' as it had become was withdrawn from Bristol Bath Road depot and broken up at Swindon in August 1972. *(Roger Holmes)*

As intimated, it was not just steam engines that were overhauled at Swindon. This is Darlington built shunter No 08 529 and originally D3691 from March 1959. When clean the blue livery could look smart but it could also weather quickly. Interestingly the 'yellow end' has been carried down on to the buffer beam whilst the 'wasp' stripes are ever present. Swindon undertook numerous shunter overhauls as BR workshops were separated under individual cost centres, they were required to bid for work with the cheapest often the winner. *(G H Taylor)*

The separate carriage works at Swindon closed in 1960 after which rolling stock repairs were concentrated within the locomotive works. Here a former Collett design brake vehicle No 4296 originally built for passenger use by the GWR, is being modified for departmental service in the Bristol Division as DW150144. Its new role being for staff accommodation engaged on tunnel inspection duties. *(G H Taylor)*

SWINDON WORKS APPRENTICE SCHOOL

The Swindon Works Apprentice School, also known as the Swindon Works Training School, took apprentices from the Works and also from the wider Western Region who would later work as maintenance engineers at depots across the WR. The new school had opened on Dean Street in 1962 and over the years taught around 2,600 students. Many of these stayed with the railway at Swindon becoming time-served and skilled artisans; others moved away although knowledge that an individual had been a 'Swindon man' held high sway at job interviews. Any number of craft skills were taught at the school and which could be useful in a chosen career. The School was designed to cater for 150 apprentices for their first year of railway experience and was fully opened on the 17th September 1962. It survived the closure of the Works and was in use until 1993.

By the 1970s work for Swindon was becoming thin on the ground. There was no new locomotive construction and rumours were rife that either Eastleigh or Swindon would close. Enter upon the scene Harry Roberts the new Works Manager from 1970, and whose pedigree included having previously been with BR Engineering Department at York (and a former WW2 Arnhem 'Red Devil'). Harry's attitude was simple, "More output from reduced space,with less staff, at the right price to the prescribed quality, within the scheduled time period."

As a start the aluminium body sheets off withdrawn diesels were melted down to produce collectors' plaques. Large work included diesel engines from locos were refurbished and sold on outside of the railway world for private use, buses were refurbished, gearboxes and engines from saw mills and other customers were rebuilt, along with missile launchers, high speed off shore boats etc. Work that could not be completed on time at other railway workshops was clandestinely brought into Swindon to be completed. This was so successful that the Works began to take on staff, and not make them redundant as in previous years. Because Swindon Works' survival was so precarious, it could never say no, and BREL knew this, so any unpopular BREL work was pushed Swindon's way. An example of this was the Advanced Passenger Train Jigs whose tight limits of construction did not appeal to the works which should have built them. Locomotive building had ceased in 1965 with the last of the (D95xx) Class 14 'Teddy Bear' diesels but 13 years later an order was placed for 20 locomotives for the railways of Kenya; and to run on metre-gauge rails.

Swindon was given a punitive time scale to build these locomotives, even though it had no equipment to do it. The other works had told the Board it was not possible to do so in the planned time-frame or for the contract price. Swindon workers would not be cowed or beaten by such advice - 'Great Western' independence lived on. 'A' shop was immediately made ready with a production line laid out and metre gauge track. This required the building of its own plant, including jigs, specialist tools etc, even knowing there would be no repeat order. It was viewed by those outside Swindon as a recipe for failure. Harry Roberts intimated that his colleagues told him he was crazy to have accepted this work for Swindon, but he had confidence in the Swindon workforce and it was justified as they turned out the locomotives on time, and on price. Other work that came their way at this time included building boats, repairing Bristol Buses, building Rail Traversers, refurbishing Vintage Cars (body shop skills), converting Bedford trucks into mobile airport plane steps, undertaking rail crash damage repairs that others could not do, servicing and repairing hydraulic shunters, building the ubiquitous BR 'BRUTE's. Swindon also restored heritage items and enabled Plasser and Theurer to build a large specialised rail machine.

In the views on these two pages are examples of shunter repairs, work on locomotives from Ford at Dagenham and a BR Departmental shunter from 1959 being thoroughly refurbished in 1981. *(Transport Treasury, Trevor Davis, and G H Taylor)*

97653
NOT TO BE MOVED

Below: A sad sight for the diesel enthusiast. A line of withdrawn 'Western' class diesels awaiting their fate at Swindon in June 1977. With regular maintenance many had years of useful life ahead. Indeed one of BR's principal freight customers Foster Yeoman Ltd enquired about the possibility of purchasing some for themselves. Fortunately a number of the class survive in preservation on heritage lines.

Right: One of the jobs Swindon took on in the 1980s was refurbishing the Kent Coast EMU stock. In many cases already 25 years old, this work which involved new windows and interiors gave the vehicles a whole new lease of life. Where once might 'Kings' and 'Castles' have stood, now stands a Southern electric unit - and there is not even a trace of a copper capped chimney, or chocolate and cream livery anywhere! *(Graham Smith)*

S9
1543
1543
1

DANGER
46
48
08 550
08 036
AXLE PAD SOAKING TANK
ALL PADS, BEFORE FITTING, TO BE
SOAKED FOR A PERIOD OF NOT LESS THAN
48 HOURS, AND AFTERWARDS LEFT TO
DRAIN FOR NOT LESS THAN 24 HOURS
OR MORE THAN 48 HOURS.

Left: Steam at Swindon in 1979. The National Railway Museum's No 6000 *King George V* receiving attention at its birthplace in 1979 - and note, minus its commemorative bell. On both sides are strange bedfellows for the King, in the form of class 08 diesel shunters which were the mainstay of work here for a time. *(Transport Treasury)*

Below: (Almost) preserved locos around the turntable at Swindon in 1977. On the left is 'Warship' No D818 *Glory* in the condition as withdrawn in 1972 but later lovingly restored to original green livery by the staff at Swindon. On the right is privately preserved 'Hymek' No D7029. The tale of *Glory* is a sad one. Swindon Works was to have been one of the major locations for the GW150 celebrations in 1985 - but then the works closure was announced. It was no surprise that the men refused to participate; who could blame them? A petition to Downing Street had no effect and in their ire D818 was simply cut-up. It might not have been steam, but it had been built at Swindon. As mentioned, today the turntable still survives but there are no locos around it. Otherwise the surviving buildings on the site have been re-purposed; for example the former Pattern Store in the structure under the water tank, now the Pattern Church. *(John Tolson)*

The stone embellishments above the entrance to the Works Offices have been a feature of Swindon for well over 150 years showing as they do Broad Gauge engines from another era. The building itself also survives and is now occupied appropriately enough by English Heritage, it is thus in safe hands. Swindon's railway history is a proud heritage, indeed once it was a tradition to have a family member who worked 'inside'. Nowadays the railway at Swindon is the main lines east, west and to Gloucester but engineering still takes place, albeit on a different site and now involved in electrification. Mr Brunel's time may have passed but his legacy and that of his Locomotive Superintendent Mr Gooch and their respective successors will live on. *(R C Riley)*